DOT HS 809 735
May 2004

Technical Report

Analysis of Crashes
Involving 15-Passenger Vans

U.S. Department
of Transportation

**National Highway
Traffic Safety
Administration**

National Center for Statistics & Analysis

People Saving People
www.nhtsa.dot.gov

Technical Report Documentation Page

1. Report No. DOT HS 809 735	2. Government Accession No.	3. Recipient's Catalog No.
4. Title and Subtitle Analysis of Crashes Involving 15-Passenger Vans		5. Report Date May 2004
		6. Performing Organization Code NRD-31
7. Author(s) Rajesh Subramanian		8. Performing Organization Report No.
9. Performing Organization Name and Address Mathematical Analysis Division, National Center for Statistics and Analysis National Highway Traffic Safety Administration U.S. Department of Transportation NRD-31, 400 Seventh Street, S.W. Washington, D.C. 20590		10. Work Unit No. (TRAIS)
		11. Contract or Grant No.
12. Sponsoring Agency Name and Address Mathematical Analysis Division, National Center for Statistics and Analysis National Highway Traffic Safety Administration U.S. Department of Transportation NRD-31, 400 Seventh Street, S.W. Washington, D.C. 20590		13. Type of Report and Period Covered NHTSA Technical Report
		14. Sponsoring Agency Code

15. Supplementary Notes

The author would like to thank Joseph Tessmer, Ph.D. of the Mathematical Analysis Division for his advice on the Logistic Regression Analysis in the report. The author would also like to thank Dennis Utter and Chou-Lin Chen of the Mathematical Analysis Division for their review and valuable comments.

Abstract

This study explores the relationship between vehicle occupancy and several other variables in the National Highway Traffic Safety Administration's (NHTSA) Fatality Analysis Reporting System (FARS) database and a 15-passenger van's risk of rollover. A univariate analysis is used to demonstrate the effect of selected variables on single-vehicle rollover crashes. Variables used include speed, number of occupants, driver experience and avoidance maneuvers. Also, a logistic regression model is constructed using data from NHTSA's State Data System – a collection of all police reported crashes for that state. The resulting model permits jointly estimating the effect of these variables on the odds and rate of rollover occurrence, conditional on being in a single-vehicle police-reported crash.

17. Key Words 15- Passenger Vans, Rollover, Occupancy Maneuvers	18. Distribution Statement Document is available to the public through the National Technical Information Service, Springfield, VA 22161 http://www.ntis.gov		
19. Security Classif. (of this report) Unclassified	20. Security Classif. (of this page) Unclassified	21. No. of Pages	22. Price

Form DOT F 1700.7 (8-72) Reproduction of completed page authorized

Table of Contents

Executive Summary

The National Highway Traffic Safety Administration's (NHTSA) National Center for Statistics and Analysis (NCSA), along with NHTSA's Vehicle Research and Test Center (VRTC), released a Research Note titled "*Rollover Propensity of 15-Passenger Vans*" in April 2001. This report combined crash data and engineering analysis to conclude that the rollover risk of 15-passenger vans increases with loading (Garrott, et al. [1]).

This Technical Report provides an in-depth analysis of crashes involving 15-passenger vans to assess the effect of occupancy level on the risk of rollover. The report is organized into two major sections, the first of which provides statistics on fatal crashes involving 15-passenger vans from 1990 to 2002 using data from NHTSA's Fatality Analysis Reporting System (FARS). The statistics in this section are for descriptive purposes only and should not be used to interpret propensity or risk of rollover in 15-passenger vans. The second section constructs a logistic regression model to model the effect of various factors, most importantly occupancy level, on the risk of rollover. The model is constructed using data from 1994 to 2001 on police-reported motor vehicle traffic crashes in Florida, Maryland, Pennsylvania, North Carolina and Utah that are part of NHTSA's State Data System (SDS). The data represent the entire cross-section of police-reported crashes and are hence more representative of the real-world experience of these vehicles.

Data from fatal crashes show that between 1990 and 2002, there were 1,576 15-passenger vans involved in fatal crashes that resulted in 1,111 fatalities to occupants of such vans. Of these, 657 vans were in fatal, single vehicle crashes, of which 349 rolled over. In 450 of these vans, there was at least one fatality, totaling up to 684 occupant fatalities in single-vehicle crashes.

A large proportion of the fatally injured van occupants were not wearing seat belts. Only 14 percent of the fatally injured occupants were properly restrained. Also, 92 percent of the belted occupants survived. About 61 percent of the occupants killed in single-vehicle crashes were ejected from the van. Proper restraining greatly reduces the chances of ejection from the van. The rate of ejection for unrestrained occupants is about 72 percent as compared to 18 percent for restrained occupants.

Single vehicle crashes are used as an exposure measure to assess the risk of rollover, as every single-vehicle crash is an opportunity for a rollover to occur. In single-vehicle crashes, the vehicle characteristics that contribute to rollover are not obscured by the effect of the forces of collision. Also, a majority of rollovers occur in single-vehicle crashes.

Analysis of data from NHTSA's State Data System reveals that the rate of rollover observed for 15-passenger vans that are loaded above half their designed seating capacity is 2.2 times the rate observed for vans loaded to or below half their capacity. This disparity is the widest among all vehicle categories. A large proportion of these high-occupancy rollovers are observed to take place on high-speed roads. However, a comparison of rates of rollover, conditional on being on a high speed road, between the two loading scenarios still show the widest disparity for 15-passenger vans.

Logistic Regression modeling of NHTSA's State data reveals that the risk of rollover in a single-vehicle crash, measured in terms of predicted odds, of vehicles loaded to their designed capacity is most elevated in the case of 15-passenger vans as compared to passenger cars, SUVs, minivans and pickup trucks. Odds are a statistical transformation of probability that is widely used to compare the chances of occurrence versus non-occurrence. This metric, directly related to parameters in the logistic regression model, neatly fits into the exercise of assessing the risk of occurrence versus non-occurrence of rollover of vehicles involved in crashes.

The odds of a rollover for a 15-passenger van at its designed seating capacity, is more than five times the odds of a rollover when the driver is the only occupant in the van. This compares to ratios of close to two for SUVs and Minivans, 1.6 for pickup trucks and 1.2 for passenger cars. This disparity in the risk of rollover between lightly loaded and fully loaded scenarios is the most significant conclusion in this report.

Speed and curved road geometry were determined to be statistically significant factors affecting rollover outcome. The odds of a rollover in high-speed roads (50+ mph) are about five times the odds in a low-speed road (Under 50 mph). The odds of a rollover on curved roads increase by two times as compared to straight roads.

High occupancy single-vehicle crashes involving 15-passenger vans are significantly fewer in number as compared to other types of vehicles. While noting the disparity in sample size and comparable overall risk of rollover, it is important to observe that there is a wider disparity in the risk of rollover between nominal and full occupancy scenarios in 15-passenger vans as compared to Passenger Cars, SUVs, Pickup Trucks or Minivans.

The conclusions in this report merely point to a higher observed rate of rollover under certain vehicle, driver and crash-related factors. The conclusions should not be misconstrued to be indicative of a specific vehicle defect or a driver-related problem.

1. Introduction

Prior Research (Garrott, et. al. [1]) has shown that fully-loaded 15-passenger vans are observed to have a higher rate of rollover as compared to lightly loaded vans. NHTSA's consumer advisory of April, 2001 was based on this research. Also, NHTSA re-issued its Consumer Advisory on the rollover propensity of these vans in April, 2002. Fifteen-passenger vans[1] are primarily used by organizations for the transportation of groups such as college sports teams, commuters, church groups, recreational groups and inmates of correctional facilities.

Fifteen-passenger vans differ from most light-trucks in that they have a larger payload capacity and the occupants sit fairly high up in the vehicle (Garrott, et. al. [1]). Loading these vans to their Gross Vehicle Weight Rating (GVWR) has an adverse effect on the rollover propensity due to the increase in center-of-gravity height. Loading the vans with passengers and cargo also moves the center of gravity rearward, increasing the vertical load on the rear tires.

This report is organized into two sections. The first section uses data from NHTSA's Fatality Analysis Reporting System (FARS). This section contains raw cross tabulations of the data to identify the circumstances surrounding **fatal crashes** involving these vans during the thirteen years from 1990 to 2002. FARS data also shows that the rate of safety belt use among occupants of 15-Passenger vans involved in fatal crashes. The use of safety belts in a rollover scenario can be a significant factor in preventing serious injury to the occupants of these vans and also prevent them from being ejected from the vehicle. It is known that fatality rates among non-ejected occupants are dramatically lower compared with the ejected occupants in the same crash (Winnicki, J. [2]).

The second section constructs logistic regression models using NHTSA's State Data System (SDS) to correlate the risk of rollover with factors related to the environment, vehicle and driver. The state data system is a database of all police-reported crashes (fatal, injury or property-damage-only crashes) in a state. Of particular interest are rollovers in single-vehicle crashes involving such vans. Single vehicle crashes are used as an exposure measure to assess the risk of rollover, as every single-vehicle crash is an opportunity for a rollover to occur. In single-vehicle crashes, the vehicle characteristics that contribute to rollover are not obscured by the effect of the forces of collision. Also, a majority of rollovers occur in single-vehicle crashes. The correlation between the loading condition (occupancy) and rollover is also presented to illustrate the adverse effect of loading on the rollover propensity of these vans.

The conclusions in this report merely point to a higher observed rate of rollover under certain vehicle, driver and crash-related factors. The conclusions should not be misconstrued to be indicative of a specific vehicle defect or a driver-related problem.

---

[1] While these vehicles actually have seating positions for a driver plus fourteen passengers, they are typically called 15-passenger vans. Also, these vehicles are actually classified as buses under 49 CFR 571.3.

Only DaimlerChrysler, Ford and General Motors manufacture vans that can be configured to seat 15 passengers. The series of vans used for this study are:

- Ford E-350 Super Duty XLT (Econoline and Club Wagons)

- Dodge Ram Van B3500/Wagon B350 (1 ton) – Discontinued in 2002.

- GMC Savanna/Rally 1-ton Extended

- Chevrolet Express 1-ton Extended

The vehicles of interest were identified in FARS and SDS using the Vehicle Identification Numbers (VINs). Although the first eleven digits of the VIN are reported in FARS, only the first seven digits of the VIN are needed to identify these vans. Although the SDS consists of data reported by seventeen states, only those states that report the VIN in their databases were included in logistic regression portion of this study. The VIN pattern and the SAS® code used to identify these vans in NCSA's FARS and SDRS database are documented in the Appendices A and B, respectively, of this report. Vehicles from all model years were included in the study.

The vans identified for inclusion in this study are the extended versions, where identifiable, of their series. Only the extended versions of the series can be configured to carry 15 passengers. However, it is conceivable that some unknown number of these vehicles left the manufacturer with seating for fewer than 15 persons, as the seating configuration/capacity is not reported in FARS, and also cannot be deciphered from the VIN. Also, there is flexibility to alter the seating capacity in such vans post-production for the purpose of carrying cargo, etc.

Figure 1 shows the number of 15-passenger vans that were registered in the U.S. as of July 1 of each year. The chart shows more than a three-fold increase in the estimated number of 15-passenger vans from 1990 to 2002. According to the figures available to NHTSA as of July 1, 2002, about 500,000 15-passenger vans were registered in the U.S. This constitutes 0.25 percent of the passenger vehicle fleet (Passenger Cars, Light trucks and Vans) in the U.S. in 2002.

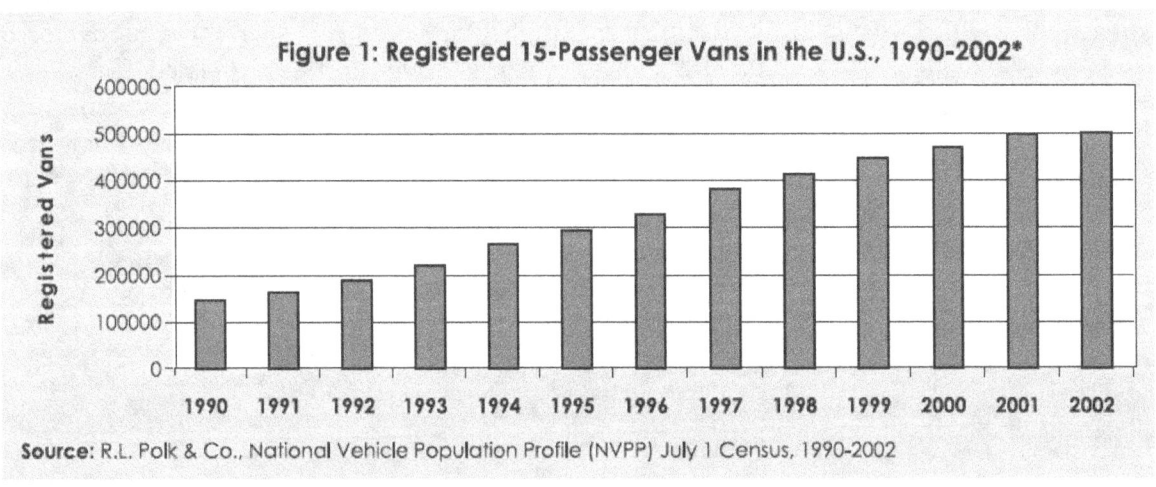

Figure 1: Registered 15-Passenger Vans in the U.S., 1990-2002*

Source: R.L. Polk & Co., National Vehicle Population Profile (NVPP) July 1 Census, 1990-2002

2. 15-Passenger Vans Involved in Fatal Crashes, 1990-2002

Data from NHTSA's Fatality Analysis Reporting System (FARS) is used in this section to present raw cross tabulations in order to identify the circumstances surrounding fatal crashes involving these vans during the twelve years from 1990 to 2002. It is important to note that fatal crash data provided in this section should not be used to interpret rollover propensity of vehicles, as the interpretation would be based on a small domain of crashes. Fatalities are a subsequent event to rollover causation where the crashworthiness of the vehicles and other factors like the use of restraints play a role in the severity of injuries.

2.1 Vehicles Involved and Fatalities

Vehicles Involved

In the period between 1990 and 2002, a total of 1,576 15-passenger vans were involved in fatal crashes resulting in 1,111 fatalities to occupants of such vans. Figure 2 shows the trend of the number of vans involved in fatal crashes.

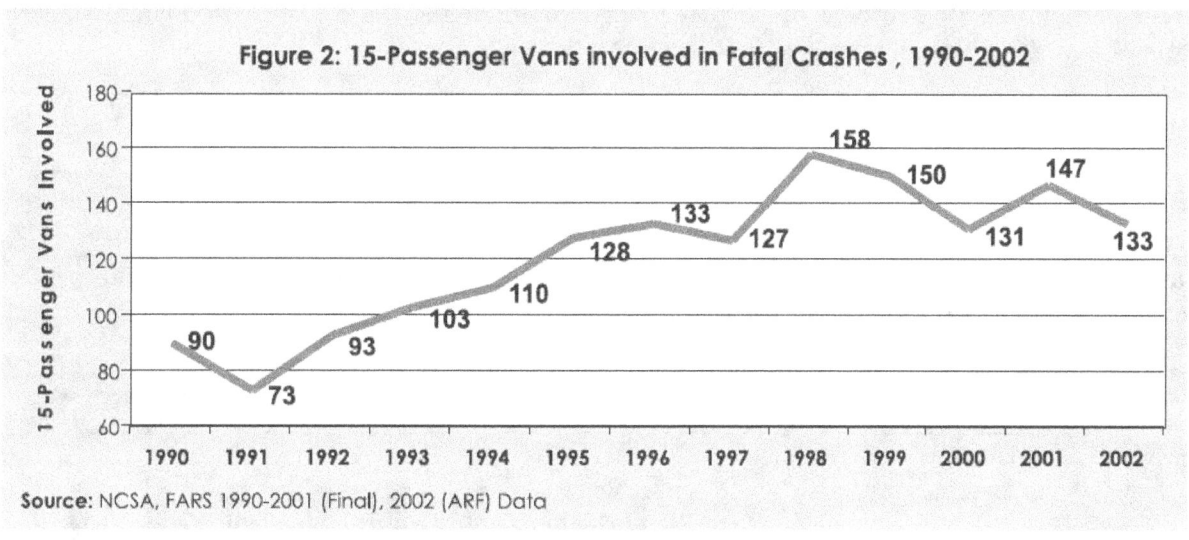

Figure 2: 15-Passenger Vans involved in Fatal Crashes , 1990-2002

Source: NCSA, FARS 1990-2001 (Final), 2002 (ARF) Data

Vehicle Involvement Rate

Figure 3 presents the vehicle involvement rate in fatal crashes per 100,000 registered 15-passenger vans.

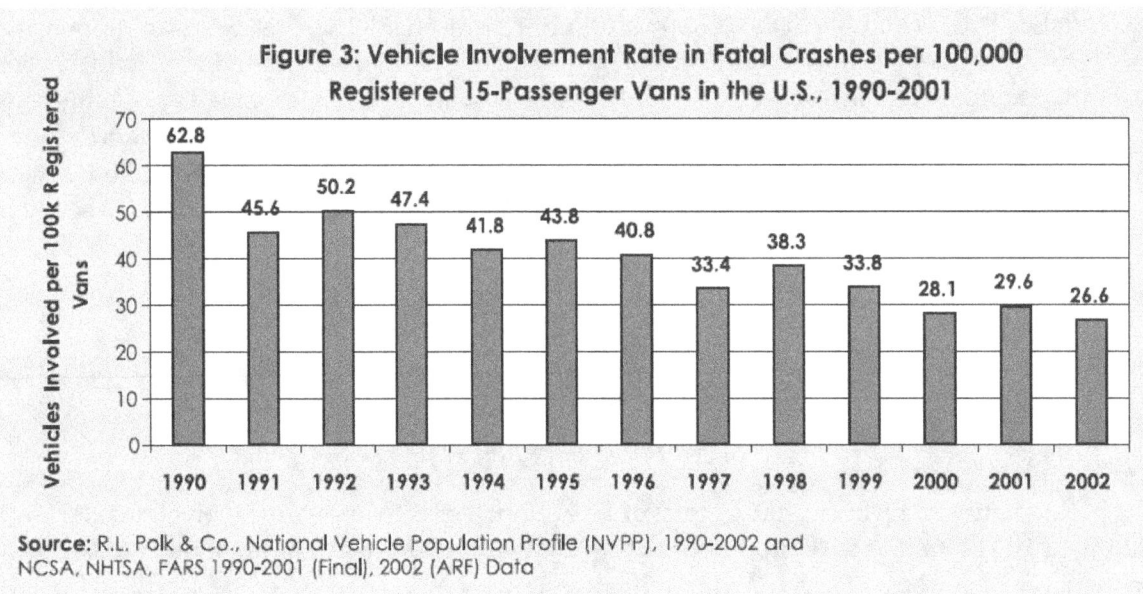

Figure 3: Vehicle Involvement Rate in Fatal Crashes per 100,000 Registered 15-Passenger Vans in the U.S., 1990-2001

Source: R.L. Polk & Co., National Vehicle Population Profile (NVPP), 1990-2002 and NCSA, NHTSA, FARS 1990-2001 (Final), 2002 (ARF) Data

In the period from 1990 to 2002, the vehicle involvement rate per 100,000 registered vans decreased from 62.8 in 1990 to an all time low of 26.6 in 2002.

Fatalities

In crashes involving the 1,576 15-passenger vans between 1990 and 2002, fatalities occurred to occupants of 15-passenger vans, occupants of other vehicles that were also involved in the crash as well as nonoccupants (pedestrians and pedalcyclists). Figure 4 illustrates the trend of fatalities by the role of the persons killed in the crash.

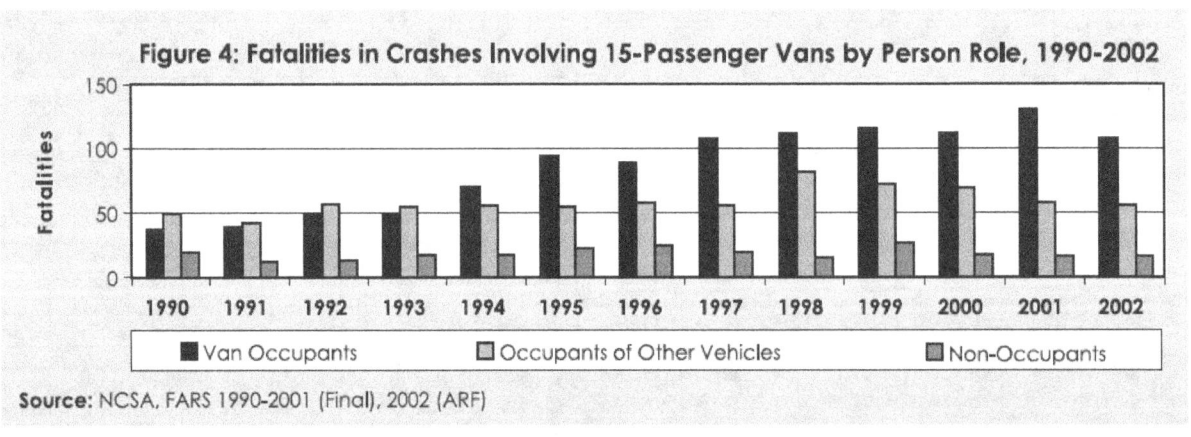

Figure 4: Fatalities in Crashes Involving 15-Passenger Vans by Person Role, 1990-2002

Source: NCSA, FARS 1990-2001 (Final), 2002 (ARF)

Table1 - Fatalities in Crashes Involving 15-Passenger Vans by Person Role, 1990-2002

Year	Van Occupants	Occupants of Other Vehicles	Non-Occupants	Total
1990	37 (35%)	50 (47%)	19 (18%)	106 (100%)
1991	39 (42%)	42 (45%)	12 (13%)	93 (100%)
1992	48 (41%)	57 (48%)	13 (11%)	118 (100%)
1993	48 (40%)	55 (46%)	17 (14%)	120 (100%)
1994	70 (49%)	56 (39%)	17 (12%)	143 (100%)
1995	94 (55%)	55 (32%)	23 (13%)	172 (100%)
1996	89 (52%)	58 (34%)	25 (15%)	172 (100%)
1997	108 (59%)	56 (31%)	19 (10%)	183 (100%)
1998	112 (54%)	82 (39%)	15 (7%)	209 (100%)
1999	116 (54%)	72 (33%)	27 (13%)	215 (100%)
2000	112 (57%)	69 (35%)	17 (9%)	198 (100%)
2001	130 (64%)	58 (28%)	16 (8%)	204 (100%)
2002	108 (60%)	56 (31%)	16 (9%)	180 (100%)
Total	1,111 (53%)	766 (36%)	236 (11%)	2,113 (100%)

Source: NCSA, FARS 1990-2001 (Final), 2002 (ARF)

Table 1 depicts the data underlying Figure 4.

As seen in Figure 4 and Table 1, about two-thirds of all fatalities in crashes involving 15-passenger vans in 2002 occurred to the occupants of the vans themselves. This proportion has increased from a low of 35 percent in 1990 to a high of 64 percent in 2001

15-Passenger Van Occupant Fatalities by Crash Type

Figure 5 breaks down occupant fatalities by the type of the crash, i.e., if the 15-passenger van was involved in a single-vehicle or multiple-vehicle crash.

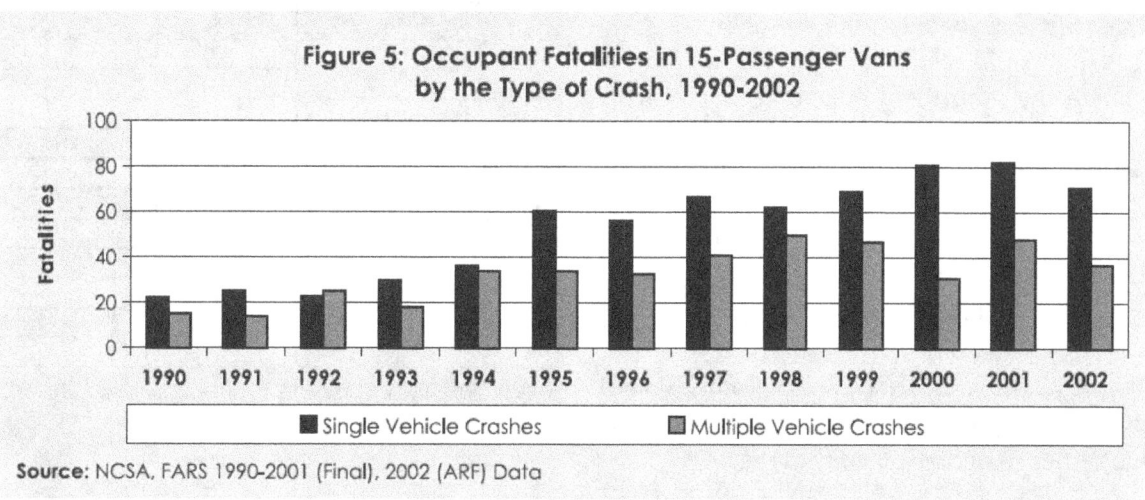

Figure 5: Occupant Fatalities in 15-Passenger Vans by the Type of Crash, 1990-2002

Source: NCSA, FARS 1990-2001 (Final), 2002 (ARF) Data

Table 2 depicts the data underlying Figure 5.

Table 2 - 15-Passenger Vans Involved in Fatal Crashes and Occupant Fatalities by Crash Type, 1990-2002

Year	Single-Vehicle*		Multiple-Vehicle		All Crashes	
	Vehicles	Fatalities	Vehicles	Fatalities	Vehicles	Fatalities
1990	32	22	58	15	90	37
1991	29	25	44	14	73	39
1992	31	23	62	25	93	48
1993	38	30	65	18	103	48
1994	42	36	68	34	110	70
1995	58	60	70	34	128	94
1996	64	56	69	33	133	89
1997	56	67	71	41	127	108
1998	59	62	99	50	158	112
1999	63	69	87	47	150	116
2000	60	81	71	31	131	112
2001	69	82	78	48	147	130
2002	56	71	77	37	133	108
Total	657	684	919	427	1,576	1,111

Source: NCSA, NHTSA, FARS 1990-2001 (Final), 2002 (ARF) Files

*Some years have more vehicles than occupant fatalities as there are crashes with no 15-passenger van occupant fatality but a pedestrian or pedal-cyclist died in the crash.

As seen in Table 2, about 62 percent (684/1,111) of fatalities to occupants of 15-passenger vans occur in single-vehicle crashes, i.e., the vans were the only vehicles involved in the crash, although there might have been pedestrians or pedal-cyclists involved in the crash. Some of the events that could result in single vehicle crashes are when the van hits a guardrail/tree or rolls over or a combination of the two. Table 3 depicts the distribution of the number of fatalities in the 1,576 vehicles involved in fatal crashes. It can be inferred from the data in Table 3 that there were 722 (1,576-854 [Table 3]) 15-passenger vans that had at least one fatally injured occupant.

Of these 722 vehicles, 450 (657-207 [Table 3]) were involved in single-vehicle crashes accounting for 684 fatalities to occupants of those vans.

Table 3 - 15-P Vans in Fatal Crashes, By Number of Fatally Injured Occupants in Van, 1990-2002

Fatalities in Vehicle	Number of Vehicles	
	All Crashes	Single-Vehicle Crashes
None	854	207
1	521	319
2	121	81
3	41	33
4	10	4
5	13	5
6	5	1
7	7	5
8	2	1
11	1	0
14	1	1
TOTAL	1,576	647

Table 4 shows the occurrence of rollover in fatal crashes and the number of van occupants that were killed in these crashes between 1990 and 2002. The rollovers shown in this table consist of all crashes for which rollover was a first or subsequent event. Rollover as a first event is coded in those crashes where the First Harmful Event in the crash was a rollover. Rollover as a subsequent event is coded in those crashes where the vehicle rolled over after an initiating first harmful event (e.g., collision with a guard-rail etc or collision with another vehicle, etc.).

Table 4 - 15-Passenger Vans Involved and Occupant Fatalities by Rollover Occurrence, 1990-2002

Year	Rollovers		No Rollovers		All Crashes	
	Vehicles	Fatalities	Vehicles	Fatalities	Vehicles	Fatalities
1990	21	23	69	14	90	37
1991	20	28	53	11	73	39
1992	25	21	68	27	93	48
1993	22	27	81	21	103	48
1994	29	42	81	28	110	70
1995	41	64	87	30	128	94
1996	41	52	92	37	133	89
1997	42	69	85	39	127	108
1998	53	71	105	41	158	112
1999	45	76	105	40	150	116
2000	55	91	76	21	131	112
2001	66	91	81	39	145	130
2002	50	70	83	38	133	108
Total	510	725	1,066	386	1,576	1,111

Source: NCSA, NHTSA, FARS 1990-2001 (Final), 2002 (ARF) Files

About two-thirds (725/1,111) of the fatalities to occupants of 15-passenger vans occurred when the vans rolled over. Table 5 shows the vehicles involved and the fatalities to occupants of these vehicles that rolled over (725 fatalities – Table 3) by the type of the crash.

Table 5 - 15-Passenger Vans that Rolled Over and Subsequent Fatalities by Type of Crash, 1990-2002

Year	Single Vehicle		Multiple Vehicle		All Rollovers	
	Vehicles	Fatalities	Vehicles	Fatalities	Vehicles	Fatalities
1990	12	17	9	6	21	23
1991	15	20	5	8	20	28
1992	15	16	10	5	25	21
1993	18	24	4	3	22	27
1994	19	25	10	17	29	42
1995	29	49	12	15	41	64
1996	31	42	10	10	41	52
1997	28	55	14	14	42	69
1998	35	49	18	22	53	71
1999	31	60	14	16	45	76
2000	39	72	16	19	55	91
2001	44	70	22	21	66	91
2002	33	57	17	13	50	70
Total	349	556	161	169	510	725

Source: NCSA, NHTSA, FARS 1990-2001 (Final), 2002 (ARF) Files

More than three-quarters (556/725) of all fatalities that occurred in rollover crashes between 1990 and 2002 were in vans involved in single-vehicle crashes. Single vehicle crashes are used as an exposure measure to assess the risk of rollover, as every single-vehicle crash is an opportunity for a rollover to occur. In single-vehicle crashes, the vehicle characteristics that contribute to rollover are not obscured by the effect of the forces of collision. Also, a majority of rollovers occur in single-vehicle crashes.

Table 6 shows the proportion of crashes when the van rolled over by the type of the crash. In 2002, 15-passenger vans involved in fatal single-vehicle crashes were more than twice as likely to have rolled over as compared to those vans that were involved in multiple-vehicle crashes. Also, about 53 percent of the 15-passenger vans involved in fatal, single-vehicle crashes rolled over. This proportion has increased from a low of 38 percent in 1990 to a high of 65 percent in 2000 and has decreased to 59 percent in 2002.

Table 6 - Proportion of 15-Passenger Vans Rollovers by Type of the Crash, 1990-2002

Year	Single-Vehicle Crashes			Multiple-Vehicle Crashes			All Crashes		
	Crashes	Rollovers	%	Crashes	Rollovers	%	Crashes	Rollovers	%
1990	32	12	38%	58	9	16%	90	21	23%
1991	29	15	52%	44	5	11%	73	20	27%
1992	31	15	48%	62	10	16%	93	25	27%
1993	38	18	47%	65	4	6%	103	22	21%
1994	42	19	45%	68	10	15%	110	29	26%
1995	58	29	50%	70	12	17%	128	41	32%
1996	64	31	48%	69	10	14%	133	41	31%
1997	56	28	50%	71	14	20%	127	42	33%
1998	59	35	59%	99	18	18%	158	53	34%
1999	63	31	49%	87	14	16%	150	45	30%
2000	60	39	65%	71	16	23%	131	55	42%
2001	69	44	64%	78	22	28%	147	66	46%
2002	56	33	59%	77	17	22%	133	50	38%
Total	657	349	53%	919	161	18%	1,576	510	32%

Source: NCSA, NHTSA, FARS 1990-2001 (Final), 2002 (ARF) Files

In the period from 1990 to 2002 about 68 percent (349/510) of all rollovers involving these vans occurred in single-vehicle crashes. Driver-related factors and vehicle dynamics (non-driver related factors) along with the influence of environmental factors can be contributing factors in a single-vehicle crash resulting in a rollover. There were 657 such crashes from 1990 to 2002 resulting in 556 fatalities of occupants of 15-passenger vans.

2.2 Restraint Use Among Occupants of 15-Passenger Vans

Fifteen-passenger vans are equipped with a safety belt (lap or lap-shoulder belt) in every seating position (driver and 14 passengers). A total of 684 occupants of 15-passenger vans were killed in a single-vehicle crash. Table 7 depicts the extent of restraint use among fatally injured occupants of 15-passenger vans in single-vehicle crashes.

As shown in Table 7, 75.6 percent of the occupants killed in fatal single-vehicle crashes were not restrained, i.e., they were not wearing safety-belts or not properly restrained in child-safety seats, etc. The chance of a serious injury is higher when an occupant is not restrained, among other things, the chances of being ejected out of the vehicle increases. Fatality rates among non-ejected occupants are dramatically lower

Table 7 - Restraint Use Among Fatally Injured Occupants of 15-Passenger Vans in Fatal, Single-Vehicle Crashes, 1990-2002

Restraint Use	Number	Percent
Unrestrained	517	75.6
Restrained	95	13.9
Unknown	72	10.5
Total	684	100.0

Source: NCSA, NHTSA, FARS 1990-2001 (Final), 2002 (ARF) Files

Table 8 - Ejection and Restraint Use Among Fatally Injured Occupants of 15-Passenger Vans in Fatal, Single-Vehicle Crashes, 1990-2002

Restraint Use	Ejection			
	Ejected	Not Ejected	Unknown	Total
Restrained	17	78	0	95
Unrestrained	371	134	12	517
Unknown	26	41	5	72
Total	414	253	17	684

Source: NCSA, NHTSA, FARS 1990-2001 (Final), 2002 (ARF) Files

compared with the ejected occupants in the same crash (Winnicki, J. [2]).

As shown in Table 8, about 72 percent (371/517) of the fatally injured, unrestrained occupants of 15-passenger vans in single vehicle crashes were ejected (partially or totally) from the van.

As seen in Table 9, an unrestrained occupant in a fatal, single vehicle crash involving a 15-passenger van is about three times as likely to have been killed as compared to a restrained occupant (22 percent versus 8 percent). The lack of data did not permit a more reasonable metric that would have been based the restraint usage rate among occupants in all crashes and the ensuing severity of injuries.

Table 9 - Injury Severity by Restraint Use Among Occupants of 15-Passenger Vans in Fatal Single-Vehicle Crashes, 1990-2002

Restraint Use	Killed	Survived	Total	p-value
Unrestrained	517 (22%)	1,816 (78%)	2,333 (100%)	<0.0001
Restrained	95 (8%)	1,055 (92%)	1,150 (100%)	<0.0001
Unknown	72 (88%)	420 (12%)	492 (100%)	-
Total	684 (17%)	3,291 (83%)	3,975 (100%)	-

Source: NCSA, NHTSA, FARS 1990-2001 (Final), 2002 (ARF) Files

Table 10 - Restraint Use Among Fatally Injured Occupants in Single Vehicle Crashes by Vehicle Type, 2002

Vehicle Type	Restrained	Unrestrained	Unknown
Passenger Cars	30	62	8
SUVs	25	70	5
Pickup Trucks	18	76	6
Vans	26	65	9
15-Passenger Vans	14	76	11

Source: NCSA, NHTSA, FARS 2002 (ARF) Files

Table 10 depicts the proportion of fatally injured occupants, in 2002, that were unrestrained by the type of vehicle that they were driving/riding in. Fatally injured occupants of 15-passenger vans and Pickup Trucks have the lowest rate of restraint use as compared to occupants of passenger cars, SUVs, and Vans.

2.3 Comparison with Other Vehicle Types Involved in Fatal Crashes

Table 11 depicts the rate of fatal, single vehicle crashes per 100,000 registered vehicles by vehicle type from 1995 to 2002 [only back to 1995 as reliable registration data exists only back to that year]. As shown in Table 6, the number of 15-passenger vans involved in fatal, single vehicle crashes per 100,000 registered vans has been decreasing since 1995, but is still higher than other categories of passenger vehicles like cars, SUVs, other vans, etc. The higher rate of fatal, single-vehicle crashes is also due to the fact that the occupancy levels in these vans are larger than those in the smaller passenger vehicles and this in turn results in a higher probability of at least one occupant fatality in the van.

Table 11 - Number of Fatal, Single-Vehicle Crashes per 100,000 Registered Vehicles, 1995-2001

Vehicle Type	1995	1996	1997	1998	1999	2000	2001
Passenger Cars	9.6	9.3	8.9	8.6	8.2	8.1	8.0
SUVs	14.3	14.1	13.6	13.3	13.0	12.7	11.8
Pickup Trucks	13.0	12.3	11.8	11.8	11.7	11.3	11.4
15-Passenger Vans	19.8	19.6	14.7	14.3	14.2	12.9	13.9
Other Vans	7.3	7.4	7.5	7.7	7.4	7.2	6.5

Source: NCSA NHTSA FARS 1990-2001 (Final) 2002 (ARF) Files R.L.Polk and Company NVPP Registration Data

3. Analysis Using Crash Data from NHTSA's State Data System (SDS)

The descriptive statistics in the previous section were based on data on fatal crashes, i.e., crashes that resulted in at least one fatally injured person. This section presents a detailed analysis of crash data from five states that are part of NHTSA's State Data System (SDS). The data are a census of all police-reported crashes in that state comprising of serious crashes (those resulting in a fatality or injury) as well as those that only resulted in damage to property. Consequently, the data are representative of the population of police-reported crashes in these states for those years.

3.1 Data and Methodology

NHTSA's state data system consists of crash data from seventeen participating states. However, not all states report the Vehicle Identification Number (VIN) that is necessary to identify 15-passenger vans. The five states that report VINs were chosen for this study. Data, spanning multiple years from these states, were included in this analysis. Table 12 depicts the states chosen and the years of data

Table 12 - States and Years of Data Chosen for Analysis

States	Years
Florida	1994 to 2001
Maryland	1994 to 2001
North Carolina	1994 to 1999
Pennsylvania	1994 to 2000
Utah	1994 to 2001

Source: NHTSA State Data Reporting System (SDRS)

included. In order to identify vehicle types (e.g., passenger cars, SUVs, 15-passenger vans, etc.), the VIN was decoded to extract vehicle model codes. These codes are stored as part of a supplemental analytic file in the SDS. The model year of the vehicle is also derived in this manner. Other variables of interest were all re-coded into a uniform variable for analysis. These variables included rollover occurrence, occupancy[1], age of the driver, driver impairment, weather conditions, roadway surface conditions, speed-limit (as a proxy for travel speed). The variables and data chosen are along the lines of those chosen for NHTSA's Rollover Assessment Program that generates star-ratings for rollover risk of passenger vehicles. Of particular interest are single vehicle crashes involving these vehicles. Single vehicle crashes are used as an exposure measure to assess the risk of rollover, as every single-vehicle crash is an opportunity for a rollover to occur. In single-vehicle crashes, the vehicle characteristics that contribute to rollover are not obscured by the effect of the forces of collision. Also, a majority of rollovers occur in single-vehicle crashes.

The results from the analysis of the state data are presented in two parts – a descriptive part outlining summary crash data by vehicle type containing rollover and crash ratios and an analytic part containing the results of a logistic regression model to predict rollover as an outcome conditional on given vehicle, driver and environmental characteristics.

---

[1] Occupancy is derived by adding up the number of occupants in the person level file. All states chosen for this analysis report all persons, injured or uninjured, involved in the crash. For this reason, the Missouri data, while fulfilling other requirements, was dropped from this analysis as not all uninjured persons are reported in the data.

3.2 Descriptive Statistics

The data in this section will describe the occurrence of crashes and rollovers by vehicle type. The major vehicle categories chosen for analysis are

- Sport Utility Vehicles (SUVs)
- Pickup Trucks (Pickups)
- Minivans
- Passenger Cars
- 15-Passenger Vans
- Other Vans
- Others/Unknown

The metric that will be used in this section, for a given crash type, is the ratio of vehicles that rolled over to number of vehicles involved in a given type of crash. This metric will be used to compare the 15-passenger van's 'propensity' to rollover as compared to that for other vehicles. At this stage, it is important to highlight the

resistance is measured by the propensity of the vehicle to roll over, conditional on a single vehicle crash having occurred.

Table 13 presents the overall picture on the number of crashes by vehicle type as reported to the six states used in this analysis.

Single-vehicle crashes, expressed as a percentage of all crashes, have low rates of incidence for 15-passenger vans as compared to other vehicle types. About 9 percent of all crashes involving 15-passenger vans were single vehicle crashes. The incidence of single-vehicle crashes as a proportion of all crashes was the lowest for Minivans (8 percent) and highest for SUVs (14 percent). So overall, it seems that 15-passenger vans do not have any unusual handling issues, which would have manifested itself in a higher incidence of single vehicle crashes, as compared to the other types of vehicles. Also, there may

Table 13 - Vehicles Involved in Crashes by Crash Type and Type of Vehicle

Vehicle Type	All Crashes	Single Vehicle Crash		Multiple Vehicle Crash	
	Vehicles	Vehicles	Percent	Vehicles	Percent
15-P Vans	15,622	1,441	9.22%	14,181	90.78%
Passenger Cars	3,625,467	423,760	11.69%	3,201,707	88.31%
SUVs	440,917	61,968	14.05%	378,949	85.95%
Pickup Trucks	752,814	98,282	13.06%	654,532	86.94%
Minivans	202,429	16,205	8.01%	186,224	91.99%
Other Vans	170,105	15,745	9.26%	154,360	90.74%
Other/Unknown	615,555	71,855	11.67%	543,700	88.33%
Total	5,822,909	689,256	11.84%	5,133,653	88.16%

Source: NHTSA State Data Reporting System (SDRS) FL, MD, NC, PA and UT data.

differences between the handling characteristics of a vehicle and its resistance to rollover. Some vehicle characteristics, such as handling problems, may result in a relatively high frequency of single vehicle crashes. The vehicle's rollover resistance can then be assessed by whether a single vehicle crash results in a rollover. The vehicle's rollover

be various driver characteristics, including some not reported/measured, may contribute to a relatively higher incidence of single-vehicle crashes. Making the analysis of rollovers conditional on being in a single-vehicle crash also captures these factors.

Table 14 depicts the incidence of rollover by vehicle type and type of crash (single or multiple vehicle). Single vehicle crashes are the preferred domain of analysis as the vehicle dynamics are more likely to have played a part in rollover causation as compared to multiple vehicle crashes, where the impact dynamics can also play a role.

Overall, the incidence of rollover in single vehicle crashes for 15-passenger vans, expressed, as a percentage of vehicles involved in such crashes, is comparable with those for other types of vehicles. SUVs had the highest incidence (39 percent) among all the vehicle categories while passenger cars had the lowest incidence rates (16 percent). However, the issue at hand is to analyze the rate of rollover at various occupancies for the different vehicle types. Prior research (Garrott, et. al) has indicated that the rate of rollover for 15-passenger vans increases three-fold when the vans have 10 occupants or more as compared to those that have fewer than 10 occupants.

Table 14 - Crashes and Rollovers by Crashes Type and Type of Vehicle

Vehicle Type	Single Vehicle Crashes			Multiple Vehicle Crashes			All Crashes		
	Crashes	Rollovers	%	Crashes	Rollovers	%	Crashes	Rollovers	%
15-P Vans	1,441	315	22	14,181	172	1	15,622	487	3
Passenger Cars	423,760	66,318	16	3,201,707	24,320	1	3,625,467	90,638	3
SUVs	61,968	23,927	39	378,949	9,625	3	440,917	33,552	8
Pickup Trucks	98,282	26,187	27	654,532	9,936	2	752,814	36,123	5
Minivans	16,205	2,746	17	186,224	2,389	1	202,429	5,135	3
Other Vans	15,745	3,592	23	154,360	2,476	2	170,105	6,068	4
Other/Unknown	71,855	14,491	-	543,700	7,943	-	615,555	22,434	-
Total	689,256	137,576	20	5,133,653	56,861	1	5,822,909	194,437	3

Source: NHTSA State Data Reporting System (SDRS) FL, MD, NC, PA and UT data.

Fully loaded conditions for the various vehicle categories are shown in Table 15. It is entirely conceivable that some individual models within a vehicle category might have a higher seating capacity than the one indicated in Table 15. Figure 6 depicts the rate of rollover in single vehicle crashes for the different vehicle types

Table 15 - Occupancies Assumed as Fully-Loaded Conditions by Type of Vehicles

Vehicle Type	Number of Occupants
15-Passenger Van	15+
Passenger Cars	4+
SUVs	4+
Pickup Trucks	4+
Minivans	7+

by occupancy. It is to be noted that in the chart, occupancy of 4 for passenger cars, pickup trucks and SUVs is actually 4 or more, occupancy of

7 is actually 7 or more occupants. It is entirely conceivable that some of the vehicles may have a designed seating capacity that exceeds those shown in Table 13. It is not possible to identify the seating configuration of passenger vehicles from NHTSA's databases or VINs. Also vehicles with much larger seating capacities than those mentioned in Table 13, especially SUVs, have been late entrants to the fleet. The latest data year in this analysis was 2001 and it is reasonable to assume that the fleet was heavily weighted towards the seating capacities mentioned in Table 13.

Figure 6 compares the rates of rollover for various vehicle types by when they are loaded to half or under their seating capacity versus over half their seating capacity. For the sake of this analysis, passenger cars, SUVs and pickup trucks with two occupants or less, minivans with three occupants or less and 15-passenger vans with seven occupants or less are defined as vehicles loaded to half their capacity or under.

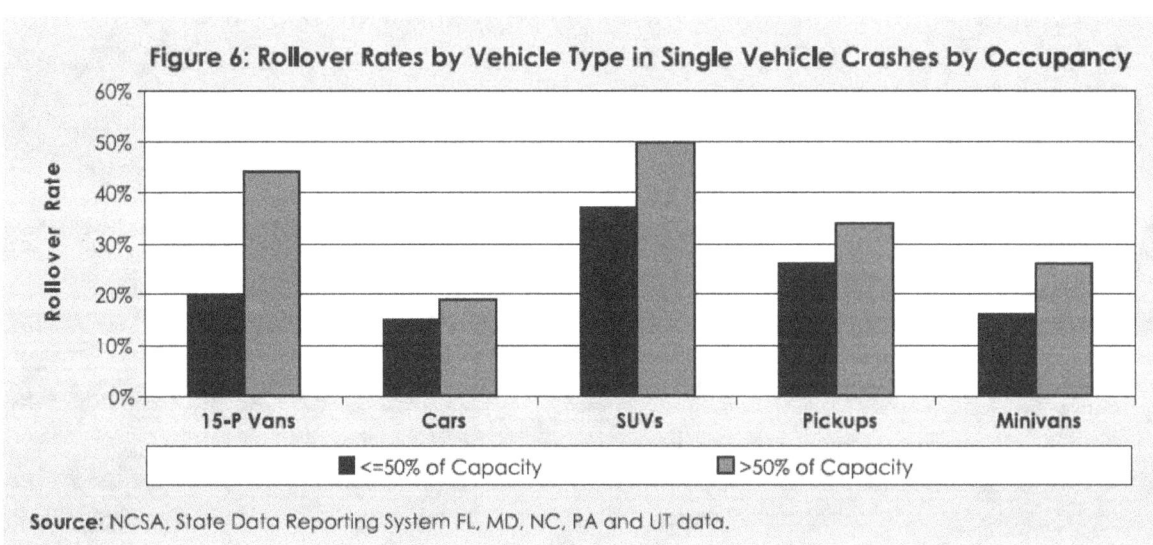

Figure 6: Rollover Rates by Vehicle Type in Single Vehicle Crashes by Occupancy

Source: NCSA, State Data Reporting System FL, MD, NC, PA and UT data.

As seen in Figure 7, when the vehicles are loaded to more than half of their seating capacity, the rates of rollover are higher as compared to when they are loaded to half their seating capacity or less. However, the relative difference in the rates of rollover under the two different loading scenarios is most pronounced for 15-passenger vans. This relative difference is shown in Table 16 for other vehicle categories. It is noted that a 15-passenger van that is loaded to half its designed seating capacity has as many occupants as any other type of passenger vehicle that is fully loaded. The differences for all vehicle categories are statistically significant, as indicated by the p-values in Table 16.

road they were traveling at the time of the crash. The percentages in each of the bars in Figure 8 indicate the proportion of the rollovers in that category that occurred on high-speed roads (50+ mph). So, 62 percent of rollovers of 15-passenger vans that loaded to half or under half of their designed capacity were in high-speed roads. In comparison, 91 percent of rollovers involving 15-passenger vans that were loaded at or above half their designed seating capacity occurred on high-speed roads.

The data in Figure 7 indicate that a great proportion of rollovers of 15-passenger vans in heavily loaded scenarios occur on high-speed

Table 16 - Rollover Rates by Occupancy and Vehicle Type in Single Vehicle Crashes

Vehicle Type	Half the Seating Capacity or Under	Over Half the Seating Capacity	Relative Difference(Ratio)	Statistical Significance
15-Passenger Van	20%	44%	2.2	p<0.0001
Passenger Cars	15%	19%	1.3	p<0.0001
SUVs	37%	50%	1.4	p<0.0001
Pickup Trucks	26%	34%	1.3	p<0.0001
Minivans	16%	26%	1.7	p<0.0001

NHTSA State Data Reporting System (SDRS) FL, MD, NC, PA and UT data.

As shown in Table 16, occupancy seems to have a pronounced effect on the rates of rollover observed in single vehicle crashes. However, there are factors other than occupancy that can have an adverse effect on a vehicles propensity to roll over. These may include the speed of travel, surface and weather conditions, experience/ training of the driver and impaired driving. The speed of travel can be a significant factor in affecting rollover outcome because greater travel speed of the vehicle provides more energy to initiate rollover. Figure 7 un-confounds the effect of speed on the proportions shown in Figure 7. In the absence of reliable measures of travel speed, the posted speed limit at the scene of the crash is used as a proxy for the speed of travel. Figure 7 shows, by vehicle type, the composition of the rollovers by occupancy and the speed limit of the

roads, as compared to other types of vehicles. It is appropriate to examine if 15-passenger vans traveling on high-speed roads, when loaded at or above half their seating capacity, have a higher risk of rollover as compared to other types of vehicles under similar circumstances. Table 17 examines this issue by comparing the rate of rollover under various combinations of speed and occupancy for the various types of vehicles involved in single-vehicle crashes. The terms lightly-loaded and heavily-loaded have been used loosely to define loading conditions above and below half the designed seating capacity.

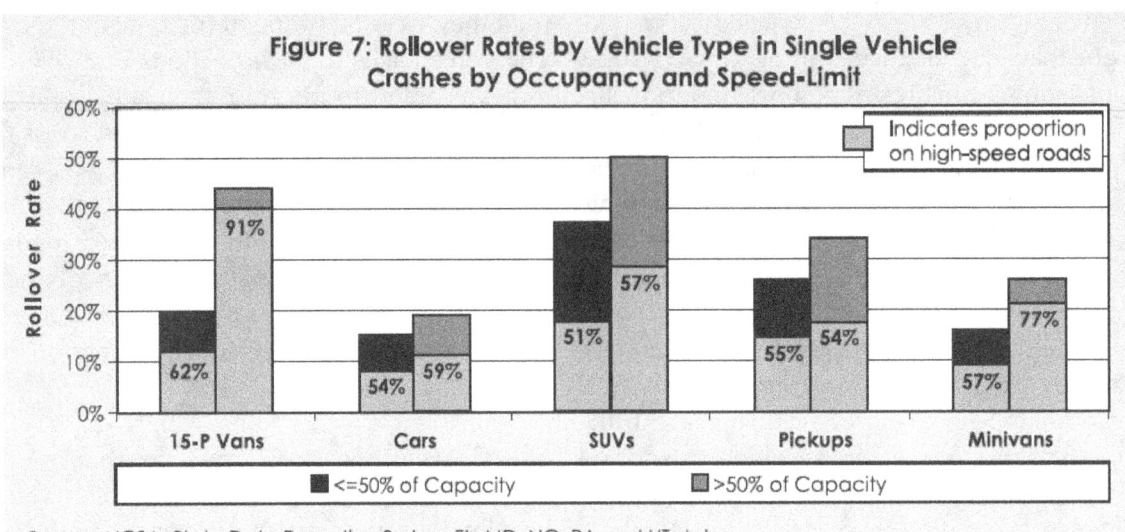

Figure 7: Rollover Rates by Vehicle Type in Single Vehicle Crashes by Occupancy and Speed-Limit

Source: NCSA, State Data Reporting System FL, MD, NC, PA and UT data.

As shown in Table 17, 15-passenger vans seem to have the highest risk of rolling over under heavily loaded scenarios in high-speed roads. Under similar circumstances, SUVs have comparable risks of rollover too. It is to be noted that the sample size of crashes for 15-passenger vans is significantly smaller than those for other types of vehicles. However, the number of crashes involving 15-passenger vans in these categories is large enough to perform a statistically valid comparison with other types of vehicles. Even though the crude rate of rollover on high-speed roads under heavily loaded scenarios for 15-passenger vans is comparable with SUVs, it is much higher than the rate for other types of vehicles. It will be noteworthy to examine the relative disparity in the rates of rollover between heavily loaded and lightly loaded scenarios on high speed roads. Table 18 depicts this relative risk ratio.

The disparity in the rates of rollover between

Table 17 - Rates of Rollover [Sample Size of Crashes] for Various Scenarios of Occupancy and Speed Limit for Vehicles Involved in Single Vehicle Crashes

Scenario	15-P Vans	Cars	SUVs	Pickup Trucks	Minivans
Lightly Loaded+ Low Speed Roads	11.7% [727]	11.5% [216,447]	29.8% [29,922]	19.6% [48,512]	11.1% [8,364]
Lightly Loaded+ High Speed Roads	29.6% [458]	22.2% [141,389]	49.2% [20,793]	35.8% [37,204]	25.6% [5,077]
Heavily Loaded+ Low Speed Roads	13.2% [38]	13.8% [3,038]	39.0% [3,379]	25.8% [2,918]	13.7% [619]
Heavily Loaded+ High Speed Roads	61.6% [86]	25.7% [4,897]	60.5% [3,565]	42.8% [2,673]	33.7% [1,050]

Source: NHTSA State Data Reporting System (SDRS) FL, MD, NC, PA and UT data.

light and heavy loading conditions on high-speed roads is the largest for 15-passenger vans. However, one can assess the true effect of occupancy on rollover propensity by taking into account the effect of various other factors that can affect rollover outcome. Statistically, a logistic regression model is very suitable to predict rollover as a dichotomous outcome (yes or no), based on explanatory variables. Logistic regression permits the joint estimation of the effect or significance of a variable in affecting rollover.

Table 18 - Rollover Rates by Occupancy and Vehicle Type in Single Vehicle Crashes in High-Speed Roads (50+ mph)

Vehicle Type	Half the Seating Capacity or Under	Over Half the Seating Capacity	Relative Difference (Ratio)
15-Passenger Van	29.6%	61.6%	2.08
Passenger Cars	22.2%	25.7%	1.16
SUVs	49.2%	60.5%	1.23
Pickup Trucks	35.8%	42.8%	1.20
Minivans	25.6%	33.7%	1.32

Source: NHTSA State Data Reporting System (SDRS) FL, MD, NC, PA and UT data.

3.3 Logistic Regression Analysis

The logit model is a regression model that is tailored to fit a dichotomous dependent variable, in this case, the occurrence or non-occurrence of rollover. The independent variables can be quantitative (e.g., number of occupants from 1 to 15+) or dichotomous (drinking/no drinking, etc.).

To appreciate the logit model, it is helpful to have an understanding of **odds** and **odds ratios**. *Probabilities* quantify the chances that an event will occur. The probability that a rollover will occur ranges from 0 to 1, with a 0 meaning that the event will almost certainly not occur, and a 1 meaning that the event will almost certainly occur. *Odds* of a rollover is the ratio of the expected number of times that an event will occur to the expected number of times it will not occur. An odds of 2 means that twice as many occurrences as non-occurrences can be expected. Similarly, an odds of ¼ means that one-fourth as many occurrences as non-occurrences are expected. So, if p is the probability of rollover and O is the odds of rollover, then:

$$O = \frac{p}{1-p} = \frac{probability\ of\ rollover}{probability\ of\ no\ rollover}$$

If the value of the odds is less than 1, the probability of rollover is below 0.5, while odds greater than 1 correspond to probabilities greater than 0.5. Like probabilities, odds have a lower

bound of 0 but there is no upper bound on odds. Odds are a more sensible scale for multiplicative comparisons. For example, if vehicle 1 is observed to have a probability of rollover of 0.30 and vehicle 2 has a probability of rollover of 0.60, then it is reasonable to claim that the probability of vehicle 2 rolling over is twice as great as the probability of vehicle 1 rolling over. However, no vehicle can have twice as much probability of rolling over as vehicle 2 (probability of 0.6x2=1.2 is not possible). On the odds scale, there are no limitations on multiplicative comparisons. A probability of 0.60 corresponds to odds of 1.5. Doubling odds of 1.5 yields odds of 3 which converts back to a probability of 0.75. This leads to the concept of *odds ratios*, a widely used measure of relationship between two dichotomous variables.

It is implicit in much of the literature on categorical data analysis that odds ratios are less sensitive to changes in marginal frequencies (e.g., the total number of rollovers and non-rollovers) than other measures of association. They are generally regarded as fundamental descriptions of the relationship between the variables of interest. Importantly, odds ratios are directly related to the parameters in the logit model.

The Logit Model

The binary response model for rollovers states that the probability of rollover, conditional on a single-vehicle crash having occurred, is a function of selected explanatory variables. It **Y** denotes the dependent variable in a binary-response model for rollovers, **Y** is equal to 1 if there is a rollover and 0 otherwise. The goal is to statistically estimate the probability that **Y=1**, considered as a function of explanatory variables. The logit model, which is a widely used binary-response model for rollover is:

$$P(Y = 1 | \mathbf{X} = x) = \frac{1}{[1 + e^{(\alpha + \beta x)}]}$$

This model can be rewritten, after taking the natural logarithm of both sides as:

$$Ln(\frac{P}{(1-P)}) = \alpha + \beta x$$

where α is the intercept and β is the vector of coefficients and x is the vector of explanatory variables. The logistic regression analysis has been performed in two ways – **independently for each vehicle type** to assess the effect of various factors in predicting rollover as well as a model for the **vehicle population as a whole** with design variables accounting for differences between the vehicle types.

3.3.1 Logistic Regression Analysis Performed Independently for Each Type of Vehicle

The explanatory variables used to model rollover as an outcome are shown in Table 19. The model uses metrics to represent various crash and driver-related characteristics and more importantly, the number of occupants in the vehicle. That is, for each vehicle type:

Logit (Pr(Rollover)) = OCCUPANCY STORM FAST HILL CURVE BADSURF MALE YOUNG OLD DRINK DUMMYFL DUMMYMD DUMMYNC DUMMYPA DUMMYUT.

The factors used in the model mirror those used in NHTSA's National Car Assessment Program (NCAP) studies.

Table 19 - Explanatory (Independent) Variables in Logistic Regression Model

Variable	Description	Levels
Occ	Number of Occupants	1 to 15+
Dark	Light Condition	1 if dark; 0 if not dark
Storm	Stormy Weather	1 if stormy; 0 if not
Fast	Speed (Speed Limit as Proxy)	1 if 50+ mph else 0
Hill	Hilly Gradient	1 if yes else 0
Curve	Road Curves	1 if yes else 0
Badsurf	Adverse Roadway Surface Conditions	1 if yes else 0
Male	Male Driver	1 if yes else 0
Young	Young Driver	1 if yes else 0
Drink	Driver Impairment	1 if yes else 0

Also included in the regression model were five variables DummyFL, DummyMD, DummyNC, DummyPA and DummyUT. The variables DUMMY<state> represent the change in Logit(Pr(Rollover)) due to the crash's taking place in that state as compared to an otherwise similar crash in Florida. They are included to control for differences in traffic patterns and reporting practices that effect rollover rates between the states. The roadway function class, i.e., if the site of the crash was a rural or urban area, was not used in the regression due to the unavailability of the data. However, it may be assumed that speed limit, curve and roadway surface conditions are reasonable explanatory variables to account for the rural/urban dichotomy. For each value of occupancy, the proportion of rollovers predicted by the model is computed by summing the predicted probabilities of rollover for all of the cases with that occupancy and dividing by the number of cases with that occupancy.

Table 16 presents the results of the logit model in terms of odds-ratios and significance parameters

$$RolloverRate_{Occupancy} = \frac{\sum_{Crashes} \Pr obabilities_{Occupancy}}{Crashes_{Occupancy}}$$

(p-values). The regression was done within each vehicle type in order to assess the effect of the various covariates on rollover outcome. Table 20 presents the odds ratios for the regression analysis on single vehicle crashes involving 15-passenger vans only.

Interpretation of Odds Ratios

Odds ratios can be interpreted as tools for multiplicative comparisons with respect to a reference value. For example, an odds ratio of 5 for *fast* indicates that the odds of a rollover on a road with a high speed limit (50 mph or above) is about five times as high as that in a lower speed road (under 50 mph), conditional on being in a single vehicle crash.

The joint estimation using the logistic regression

occupancy increases the odds of a rollover by close to 12 percent (from 1 to 1.120 or 12 percent), conditional on being in a single-vehicle crash.

Also, the odds of a rollover on a road with a high speed limit (**FAST:** 50 mph or above) is about five times as high as that in a lower speed road, conditional on being in a single vehicle crash. The odds of a rollover on a curved road (**CURVE**) increase by 99 percent over the odds of rolling over on a straight road.

Table 20 - Logistic Regression Predicting Rollover in Single-Vehicle Crashes, 15-Passenger Vans

Variable	Odds Ratio	Coefficient	P-value (Chi Sq.)
Intercept	NA	-2.1178	< 0.0001
Occupancy	1.120	0.1135	<0.0001
Dark	1.200	0.1821	0.2944
Storm	1.351	0.3011	0.1468
Fast	5.022	1.6138	<0.0001
Hill	1.087	0.0836	0.6368
Curve	1.989	0.6874	0.0001
badsurf	0.980	-0.0207	0.9237
Male	1.036	0.0349	0.8486
Young	1.222	0.2005	0.2819
Old	0.845	-0.1687	0.7613
Drink	0.811	-0.2091	0.5205

p-value for entire model: <0.0001
Source: NHTSA State Data Reporting System (SDRS) FL, MD, NC, PA and UT data.

model reveals that the variables with the most significant impact on rollover outcome, as indicated by their p-values, are:

- **Fast** (high-speed road)
- **Occupancy** (number of occupants in the vehicle)
- **Curve** (curved geometry at site)

The effect of these factors on rollover outcome is also statistically significant as indicated by the low p-values.

As seen in Table 20, each unit increase in

A comparison of the odds-ratio estimates of the three statistically significant factors (Speed, occupancy, adverse weather and road geometry) for 15-passenger vans with the corresponding odds-ratios for other vehicle types is illustrated in Table 21.

As seen in Table 21, the three factors that were significant in predicting rollover of 15-passenger vans in single vehicle crashes were also significant for other types of vehicles. For 15-Passenger vans, a unit increase in occupancy, controlling for other factors, contributes to a 12 percent (odds ratio of 1.120) increase in the predicted odds of rollover, conditional on being in a single vehicle crash. An odds-ratio of 1.12 for occupancy nature of odds ratios for different increments of occupancy. If O is the odds ratio for a unit occupancy, then the odds ratio for k occupants is O^k. Correspondingly, when loaded to the design capacity of 15 occupants, the odds ratio would be 5.47 [1.12^{15}]. Correspondingly, when passenger cars are loaded to their capacity, the odds ratio

Table 21 - Odds-Ratio Estimates of Occupancy, Road Curvature and Speed in Logit Model Predicting Rollover in Single-Vehicle Crashes, by Vehicle Type

Vehicle Type	Variable	Odds Ratio	P-value (Chi Sq.)
15-Passenger Vans	Occupancy	1.120	<0.0001
	Fast	5.022	<0.0001
	Curve	1.989	0.0001
Passenger Cars	Occupancy	1.061	<0.0001
	Fast	2.454	<0.0001
	Curve	1.889	<0.0001
SUVs	Occupancy	1.211	<0.0001
	Fast	2.626	<0.0001
	Curve	1.605	<0.0001
Pickup Trucks	Occupancy	1.134	<0.0001
	Fast	2.669	<0.0001
	Curve	1.827	<0.0001
Minivans	Occupancy	1.123	<0.0001
	Fast	3.213	<0.0001
	Curve	1.664	<0.0001

Source: NHTSA State Data Reporting System (SDRS) FL, MD, NC, PA and UT data.

implies that for every unit increase in occupancy, the odds of a rollover are increased 1.12 times – an increase of 12 percent. In order to determine the effect of increasing occupancy on rollover, it is helpful to understand the multiplicative

Table 22 - Odds-Ratio Estimates at Full Occupancy, by Vehicle Type

15-Passenger Vans (15+)	5.47
Passenger Cars (4+)	1.27
SUVs (4+)	2.15
Pickup Trucks (4+)	1.65
Minivans (7+)	2.01

Source: Logistic Regression on FL, MD,NC, PA and UT Data.

increases to 1.27 (1.061^4), or, just a 27 percent increase. Table 22 depicts these comparisons by vehicle type. The odds ratio at the designed seating capacity show the most pronounced effect for 15-passenger vans followed by Minivans, SUVs, Pickup Trucks and Passenger Cars.

In terms of change in the odds of rollover per unit increase in occupancy, 15-passenger vans compare on the same scale as other types of vehicles. However, the large multiplicative factor in terms of the number of occupants correspondingly predict much higher odds of rollover at designed seating capacity as compared to other types of vehicles. In fact, they have about 2.7 [5.41/2.01] times the odds ratio of rollover as compared to minivans at full occupancy, which is about half the capacity of 15-passenger vans. Also, 15-passenger vans have an estimated odds ratio of 4.3 times [5.47/1.27] that of passenger cars and pickup trucks when loaded to the designed seating capacity. The corresponding ratio when compared with SUVs is about 2.54 [5.47/2.15].

For the sake of comparison with Minivans, at occupancy level of 7, the odds-ratio for 15-passenger vans is 2.21 [1.12^7] – pointing to a two-fold increase in the odds of rollover. Similarly, at an occupancy level of 4, the odds-ratio for 15-passenger vans is 1.57 [1.12^4] – pointing to a 57 percent increase in the odds of rollover.

3.3.2 Logistic Regression Analysis Performed for the Vehicle Population as a Whole

The explanatory variables used to model rollover as an outcome are shown in Table 23. The model uses metrics to represent various crash and driver-related characteristics and more importantly, the number of occupants in the vehicle.

Logit (Pr(Rollover)) = OCCUPANCY STORM FAST HILL CURVE BADSURF MALE YOUNG OLD DRINK DUMMYMD DUMMYNC DUMMYPA DUMMYUT **D_CAR D_SUV D_ PICKUP D_MINIVAN**

This model will facilitate a comparison between the different vehicle types after adjusting for all other factors, including occupancy. The design variables D_CAR D_SUV D_PICKUP and D_ MINIVAN will account for overall differences in the geometry and features by the type of the vehicle. It is to be noted that the design variables average out the differences that might exist within a vehicle type, for example a compact sedan versus a large passenger car. This type of analysis is meant to provide insight into differences that might exist between different vehicle types in an overall sense and should not be interpreted for individual vehicle models within a vehicle category.

Table 23 - Explanatory (Independent) Variables in Logistic Regression Model

Variable	Description	Levels
Occ	Number of Occupants	1 to 15+
Dark	Light Condition	1 if dark; 0 if not dark
Storm	Stormy Weather	1 if stormy; 0 if not
Fast	Speed (Speed Limit as Proxy)	1 if 50+ mph else 0
Hill	Hilly Gradient	1 if yes else 0
Curve	Road Curves	1 if yes else 0
Badsurf	Adverse Roadway Surface Conditions	1 if yes else 0
Male	Male Driver	1 if yes else 0
Young	Young Driver	1 if yes else 0
Drink	Driver Impairment	1 if yes else 0
D_CAR	Design Variable for Cars	1 is Passenger Car else 0
D_SUV	Design Variable for SUVs	1 is SUV else 0
D_PICKUP	Design Variable for Pickups	1 is Pickup else 0
D_MINIVAN	Design Variable for Minivans	1 is Minivan else 0

The logistic regression yields the parameter estimates and odds ratios for the various factors as shown in Table 24.

Table 24 - Logistic Regression Predicting Rollover in Single-Vehicle Crashes, All Vehicles

Variable	Odds Ratio	Coefficient	P-value (Chi Sq.)
Intercept	NA	-1.8774	< 0.0001
Occupancy	1.095	0.0905	<0.0001
Dark	0.967	-0.0335	<0.0001
Storm	0.861	-0.1500	<0.0001
Fast	2.605	0.9574	<0.0001
Hill	1.141	0.1319	<0.0001
Curve	1.844	0.6119	<0.0001
badsurf	0.877	-0.1318	<0.0001
Male	0.995	-0.00534	0.4986
Young	1.344	0.2957	<0.0001
Old	0.667	-0.4056	<0.0001
Drink	1.266	0.2362	<0.0001
D_CAR	0.654	-0.4239	<0.0001
D_SUV	2.405	0.8777	<0.0001
D_PICKUP	1.261	0.2322	<0.0001
D_MINIVAN	0.817	-0.2016	<0.0001

p-value for entire model: <0.0001
Source: Logistic Regression on FL, MD, NC,PA and UT Data.

Plugging the coefficients, the logistic regression model yields predicted probability of rollover as shown in Figure 8. Figure 8 represents the probability distribution of rollover, conditional on a single vehicle crash, for what can be considered as a "best-case" scenario in terms of factors that affect rollover as an outcome. The "favorable" scenario is a combination of favorable driving conditions and factors for the terms included in the logistic regression model. This includes good light and weather conditions, low-speed road (under 50 mph), flat terrain, straight and good road conditions and no driver impairment.

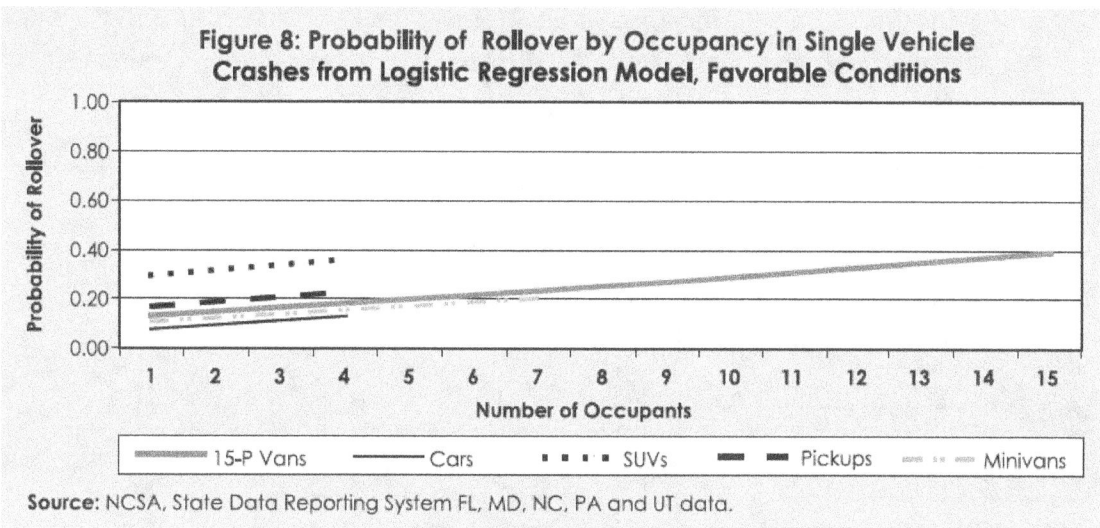

Figure 8: Probability of Rollover by Occupancy in Single Vehicle Crashes from Logistic Regression Model, Favorable Conditions

Source: NCSA, State Data Reporting System FL, MD, NC, PA and UT data.

Figure 9 depicts the distribution of the probability of rollover for what can be considered as a "adverse" scenario to affect rollover. The adverse scenario includes statistically significant variables, **fast** and **curve**. The probabilities depicted in Figure 9 are for crashes occurring on curved areas on high-speed roads.

As seen in Figures 8 and 9, the probability of rollover as indicated by the logistic regression model

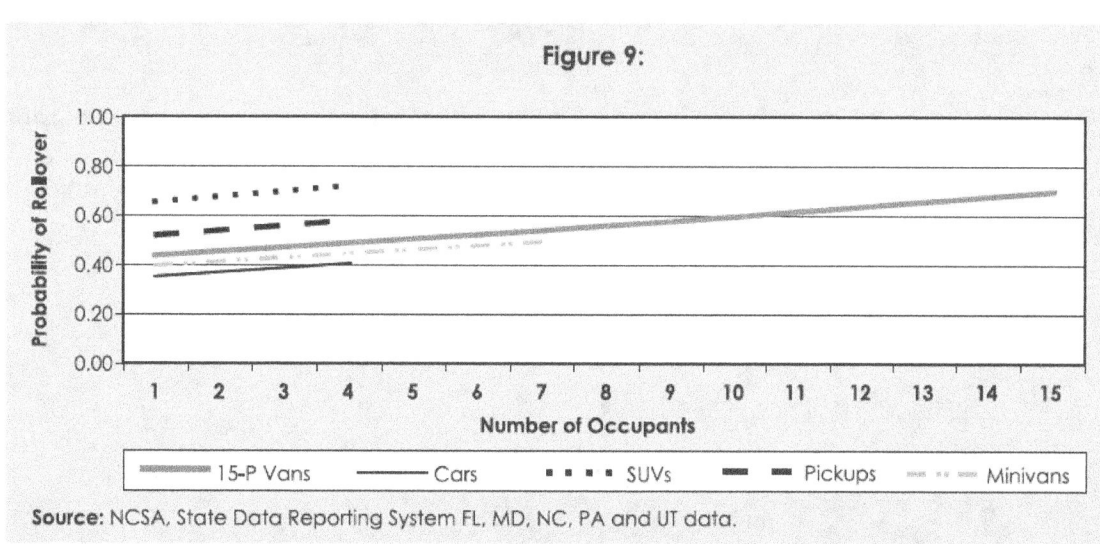

Figure 9:

Source: NCSA, State Data Reporting System FL, MD, NC, PA and UT data.

indicates a progressively worsening risk of rollover with increasing occupancy for all vehicle types including 15-passenger vans. The probability of rollover with just the driver in the vehicle ranges from under 0.20 in favorable conditions to above 0.4 in adverse conditions. However, when the van is loaded to or above its designed seating capacity, the corresponding probabilities increase to an estimated 0.40 and 0.80, respectively. This trend, while observed for all types of vehicles, is most pronounced for 15-passenger vans because of the sheer multiplicative effect of the larger seating capacity for 15-passenger vans. Figure 10 depicts various regression curves that depict how the probability of rollover in single vehicle crashes involving 15-passenger vans change upon the addition of various adverse factors that can be considered to affect rollover as an outcome.

Figure 10 depicts the relative shifts in the estimated probabilities of rollover of a 15-passenger van by

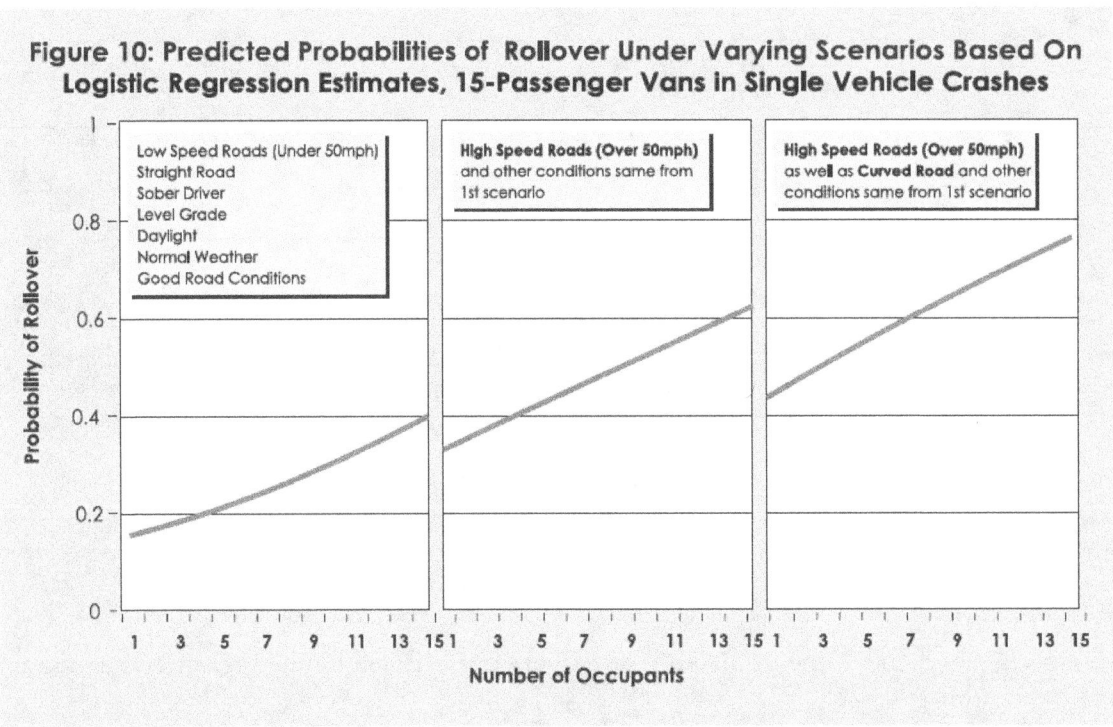

Figure 10: Predicted Probabilities of Rollover Under Varying Scenarios Based On Logistic Regression Estimates, 15-Passenger Vans in Single Vehicle Crashes

occupancy, conditional on being in a single vehicle crash, for various combinations of scenarios that could affect rollover outcome. As expected, the probability curves have higher starting values (at Occupancy=1) for more adverse scenarios but follow a more or less progressively worsening rollover rate with increasing

7. Conclusions

The purpose of this report was to analyze the circumstances in crashes that resulted in the rollover of a 15-passenger van. State crash data from five states in the period from 1994 to 2001 were used to identify vehicle, environmental and driver related factors that were significant in affecting rollovers in single vehicle crashes involving these vans. Of particular interest was the effect on increasing occupancy on the rollover propensity of 15-passenger vans. For comparison, similar analyses were performed for other types of passenger vehicles (SUVs, Pickup Trucks, Minivans and Passenger Cars). Also, NHTSA's FARS data from 1990 to 2002 were used to examine the circumstances in fatal, rollover crashes involving 15-passenger vans. The extent of seat belt use among fatally injured occupants of 15-passenger vans was compared with that of occupants of other types of vehicles. While comparisons can be made of restraint use among fatally injured occupants of different vehicle types, the true extent of belt usage can only be assessed by analyzing data that is representative of all crashes – data that is not presently available.

The overall rate of rollover in single vehicle crashes, as observed from NHTSA's State Data, for 15-passenger van is in fact lower than that for SUVs and Pickup Trucks. However, the effect of occupancy is observed to have a wider disparity in rollover rates in 15-passenger vans between conditions when the vehicle was loaded above or below half its designed seating capacity. In fact, when loaded to above half their seating capacity, 15-passenger vans were observed to have 2.2 times the rollover rate as compared to when they were loaded to or below half their designed seating capacity. This compares to lower ratios for SUVs (1.4), Pickup Trucks (1.3), Passenger Cars (1.3) and Minivans (1.7).

A majority of the high occupancy level rollovers involving 15-passenger vans were in high-speed roads (50+ mph). However, conditional on a crash having occurred on a high-speed road, the disparity in the rollover rates between scenarios when they were loaded to or below half their seating capacity and when they were loaded above half the designed seating capacity was most pronounced for 15-passenger vans. The rollover rate under the heavily-loaded scenario was again more than 2 times the rate under the lightly loaded scenario. This compares to lower ratios for SUVs (1.23), Pickup Trucks (1.20), Passenger Cars (1.16) and Minivans (1.32).

The overall rate of rollover, expressed as the proportion of single vehicle crashes that resulted in a rollover, for 15-passenger vans is, if not lower, comparable with that of other passenger vehicles. However, statistical analysis based on state crash data shows that 15-passenger vans exhibit much higher risk, measured in terms of odds, of rollover when the vans were traveling at their full capacity as compared to when the driver was the only occupant in the van. While the increment in the risk of rollover with every unit increase in occupancy for 15-passenger vans was comparable to other passenger vehicles, 15-passenger vans exhibited a much higher risk of rollover when they were loaded at or above their designed seating capacity. In fact, the odds of rollover for 15-passenger vans with 15 or more occupants was more than five times the risk of rollover when the driver was the only occupant in the van. This increase in the risk of rollover at or above the designed seating capacity is much less for SUVs and Pickup Trucks (about 2 times), Minivans (1.7 times) and Passenger Cars (1.3 times). In summary, while the overall rollover rate for 15-passenger vans are comparable to other passenger vehicles, the disparity in the risk of rollover under fully and lightly loaded conditions is most pronounced for 15-passenger vans because they can carry a larger number of occupants.

The analysis also showed speed and the geometry of the road to be factors significant in affecting rollover outcome in all types of vehicles. The posted speed limit, used as an explanatory variable for travel speed, was determined to have a significant effect on the risk of rollover for 15-passenger vans as compared to other types of vehicles. In fact, the risk of rollover, as measured by the odds, for a 15-passenger van that is traveling on a high-speed road (50+ mph) is about five times the risk of rollover for a van that is traveling on a low-speed road (under 50 mph).

The geometry of the road, as in if the road is curved or not, also was found to have a significant role in affecting rollover outcome in 15-passenger vans. The risk of rollover, measured by the odds, for a 15-passenger van traveling on a curved road is about twice the risk of rollover for a van that is traveling on a straight road.

Analysis of FARS data showed that in the period from 1990 to 2002, there were 1,111 fatally injured occupants of 15-passenger vans. Of these fatalities, 684, or about 60 percent of all 15-passenger van occupant fatalities, occurred in single vehicle crashes. Slightly more than 80 percent of all fatalities in single vehicle crashes involving 15-passenger vans occurred when the vans rolled over.

The observed safety belt usage rate is very low among fatally injured occupants of 15-passenger vans involved in single-vehicle crashes. More than three-fourths of 15-passenger van occupants killed in single-vehicle crashes were not properly restrained. Also, a majority (92 percent) of those who were properly restrained survived the crash. An unrestrained 15-passenger van occupant involved in a fatal, single vehicle crash is about three times as likely to have been killed as compared to a properly restrained occupant.

Proper restraint use greatly reduces the chances of ejection from a 15-passenger van. About 60 percent of the 15-passenger van occupants killed in single-vehicle crashes were ejected from the vehicle. An unrestrained occupant of a 15-passenger van is about four times as likely to be ejected from the van as compared to a properly restrained occupant.

Outreach efforts on this topic should emphasize the significant disparity between the risks of rollover of a 15-passenger van between lightly loaded (driver only) and fully loaded (15+ occupants) conditions. Drivers of 15-passenger vans ought to be cognizant of this change in risk when they are driving a van that is fully loaded. They also should be driven with utmost care while driving on high-speed roads as well as while negotiating a turn – conditions shown to have a significant impact in increasing the risk of rollover in any vehicle. Also, all occupants should be properly restrained when the vehicle is in motion to reduce the risk of occupant ejection in rollover events. Also, driver training on safe operation of these vans, especially of fully-loaded ones traveling on high-speed roads, is recommended.

8. References

1. Garrott, R.W., (2001) *The Rollover Propensity of Fifteen-Passenger Vans*, National Highway Traffic Safety Administration, Department of Transportation

2. Winnicki, J., (1996) *Estimating the Injury-Reducing Benefits of Ejection-Mitigating Glazing*, DOT HS 808-369, National Highway Traffic Safety Administration, Department of Transportation

15-Passenger Van: Vans that have seating positions for a driver plus fourteen passengers are typically called 15-passenger vans. Also, these vehicles are actually classified as buses under 49 CFR 571.3.

Commercial Drivers License (CDL): As of April 1, 1992, a CDL is required by all states for driving a commercial motor vehicle in excess of 26,000 pounds; or for transporting hazardous materials in sufficient amounts to be placarded; or for transporting 16 or more passengers, including the driver.

Crash BAC: The highest BAC among all the actively-involved persons in the crash. For example, in a crash involving a vehicle and a pedestrian, if the driver of the vehicle had a BAC of 0.01 g/dl and the pedestrian had a BAC of 0.11 g/dl, the Crash BAC is 0.11 g/dl.

Driver BAC: The BAC of any driver involved in a crash.

Fatal Crashes: A motor vehicle traffic crash where there was at least one fatality to a vehicle occupant(s) or nonoccupant(s) involved in the crash.

FSVC: Fatal Single-Vehicle Crash (FSVC) is a single-vehicle crash that resulted in a fatality to the occupants of the vehicle and (or) a nonoccupant that was hit by the vehicle.

Impaired: A person is said to be impaired if their BAC is between 0.01 and 0.07 g/dl (0.01-0.07).

Intoxicated: A person is said to be intoxicated if their BAC is 0.08 g/dl or greater (0.08+).

Light Trucks: A Pickup truck or a Sport Utility Vehicle (SUV).

Multiple-Vehicle Crash: A motor vehicle traffic crash is said to be a multiple-vehicle crash if more than one vehicle-level Police Accident Reports (PARs) are submitted for the crash. A multiple-vehicle crash may also involve more than one vehicle involved with one or more nonoccupants.

Nonoccupant: Any person involved in a crash who is not the occupant of a motor vehicle. Pedestrians, pedalcyclists, persons on roller-blades, skateboards, etc. are nonoccupants.

Passenger Vehicle: A passenger car, light-truck or van.

Restraint Usage: Restraints can be in the form of safety belts, child safety seats or helmets. Safety-belts are present in every seating position in a 15-Passenger Van. Restraint-Use would be coded as 'Yes' for occupants of a 15-passenger van if the use of safety belts or child safety seats is observed.

Rollover: A rollover is defined as any vehicle rotation of 90 degrees or more, about any true longitudinal or lateral axis. Rollover can occur at any time during the unstable situation. Rollover can occur as a First or Subsequent event. Subsequent event refers to a rollover that occurs after the first harmful event.

Single Vehicle Crash: A motor vehicle traffic crash is said to be a single-vehicle crash if only one vehicle-level Police Accident Report (PAR) is submitted for the crash. A single-vehicle crash can also be a crash where one vehicle was involved with one or more nonoccupants.

Appendix A: VIN Pattern to Identify 15-Passenger Vans

The vans identified for inclusion in this study are the extended versions, where identifiable, of their series. Only the extended versions of the series can be configured to carry 15 passengers. However, it is conceivable that some unknown number of these vehicles left the manufacturer with seating for fewer than 15 persons as the seating configuration/capacity is not reported in FARS and also cannot be deciphered from the VIN. Also, there is flexibility to alter the seating capacity in such vans post-production for the purpose of carrying cargo etc.

Only the first eleven positions in the VIN are needed to identify these vans as shown in Table A-1. Wildcards (*) in any of these positions indicate any character while ranges of numbers indicate all acceptable numbers for that position (e.g., 1-5 indicates 1,2,3,4,5). Character choices such as 'B/D' indicate that either 'B' or 'D' is acceptable for that position.

Table A-1 - VIN-Pattern to Identify Vans Used in This Study

Manufacturer	Position in the Vehicle Identification Number (VIN)										
	1	2	3	4	5	6	7	8	9	10	11
Ford	1-5	B/F	B/D	*	S	3	1	*	*	*	*
Chrysler	1-3/J	B	5-6	*	B	3	*	*	*	*	*
Chevrolet	1-4/J	C/G	A/B	*	G	3/8	9	*	*	*	*
GMC	1-4/J	C/G	J/D	*	G	3/8	9	*	*	*	*

Source: NCSA

Appendix B: SAS® Code to Identify 15-Passenger Vans in FARS

```
DATA VANS;
SET FARS.VEHICLE;

/*      EXTRACT FORD VEHICLES    */
IF SUBSTR(VIN,1,1) IN ('1','2','3','4','5') AND
        SUBSTR(VIN,2,1) IN ('B','F') THEN DO;
        IF SUBSTR(VIN,5,3) IN ('S31') THEN DO;
                IF SUBSTR(VIN,3,1) IN ('B','D') THEN DO;
                        VTYPE=1;
                        MODL='FORD';
                        OUTPUT;
                END;
        END;
END;

/*      EXTRACT DODGE VEHICLES */

IF SUBSTR(VIN,1,1) IN ('1','2','3','J') AND SUBSTR(VIN,2,1) IN ('B') THEN DO;
IF SUBSTR(VIN,5,2) IN ('B3') AND SUBSTR(VIN,3,1) IN ('5','6') THEN DO;
        VTYPE=1;MODL='DODG';OUTPUT;
END;
END;

/*      EXTRACT GM VEHICLES       */

IF SUBSTR(VIN,1,1) IN ('1','2','3','J','4') AND SUBSTR(VIN,2,1) IN ('C','G') THEN DO;
        IF SUBSTR(VIN,3,1) IN ('A','B') AND SUBSTR(VIN,5,3) IN ('G39','G89') THEN DO;
                VTYPE=1;MODL='CHEV';OUTPUT;
        END;
END;
IF SUBSTR(VIN,1,1) IN ('1','2','3','J','4') AND SUBSTR(VIN,2,1) IN ('C','G') THEN DO;
        IF SUBSTR(VIN,3,1) IN ('J','D') AND SUBSTR(VIN,5,3) IN ('G39','G89') THEN DO;
                VTYPE=1;MODL='GMC ';OUTPUT;
        END;
END;
RUN;
```

Appendix C: Useful Resources

1. **NHTSA Reports and Research Notes**

    *   Garrott, R.W. et al., (2001) *The Rollover Propensity of Fifteen-Passenger Vans*, National Highway Traffic Safety Administration, Department of Transportation.

    *   Subramanian, R. (2004), *Analysis of Crashes Involving 15-Passenger Vans*, National Highway Traffic Safety Administration, Department of Transportation.

2. **Safety Flyers and Hangtags**

    *   Safety Flyer: *Reducing the Risk of Rollover Crashes in Fifteen Passenger Vans*, National Highway Traffic Safety Administration, Department of Transportation.
        http://www.nhtsa.dot.gov/cars/problems/studies/15PassVans/Index.htm

    *   Hangtag: *Reducing the Risk of Rollover Crashes in Fifteen Passenger Vans*, National Highway Traffic Safety Administration, Department of Transportation.
        http://www.nhtsa.dot.gov/cars/problems/studies/15PassVans/ROLLOVER HANGTAG_LaserRes.pdf

        ### NHTSA Contact:
        Mark Krawczyk
        Consumer Programs Analyst, Consumer Information Division
        NHTSA

        Mark.Krawczyk@nhtsa.dot.gov

        (202) 366-6330

3. **Training and Instructional Materials** (NHTSA does not endorse this product but is merely providing information of its existence).

    *   ***Coaching the Van Driver*, Training Video:**

        National Safety Council   FLI Learning Systems
        (800) 621-7619   (609) 466-9000
        http://www.nsc.org/   http://www.flilearning.com/

4. **Federal Law Prohibiting Sale/Lease of 12/15-Passenger Vans for Significant Transport of Kids from School/Day Care**

    *   *NHTSA Interpretation Letter:*

        http://www.nhtsa.dot.gov/cars/rules/interps/files/17730.drn.htm

DOT HS 809 729
May 2004

www.ingramcontent.com/pod-product-compliance
Lightning Source LLC
Chambersburg PA
CBHW081624170526

45166CB00009B/3089